The Galena Nuclear Project

The Galena Nuclear Project
Pursuing Low Cost Energy in Bush Alaska

Marvin L. Yoder

Copyright © 2014 by Marvin L. Yoder.

Library of Congress Control Number:		2014908792
ISBN:	Hardcover	978-1-4990-2031-1
	Softcover	978-1-4990-2034-2
	eBook	978-1-4990-2027-4

All rights reserved. No part of this book may be reproduced or transmitted in any form or by any means, electronic or mechanical, including photocopying, recording, or by any information storage and retrieval system, without permission in writing from the copyright owner.

Any people depicted in stock imagery provided by Thinkstock are models, and such images are being used for illustrative purposes only.
Certain stock imagery © Thinkstock.

This book was printed in the United States of America.

Rev. date: 05/16/2014

To order additional copies of this book, contact:
Xlibris LLC
1-888-795-4274
www.Xlibris.com
Orders@Xlibris.com
611828

CONTENTS

Galena, Alaska 13

Infrastructure Needs 19

A Chance Meeting 26

The Toshiba 4S 29

Is This for Real? 33

Using the Energy 41

Nuclear Concerns 47

Meeting Reality 51

Final Thoughts 58

Appendix 63

Dedication

Dedicated to Dr. Norihiko Handa

Handa-san was a visionary who recognized small nuclear technology could provide low cost, reliable energy and other infrastructure in economically disadvantaged communities.

FOREWORD

It is a rare event when a single person with a passionate idea can change the course of a multi-billion dollar enterprise and set an entire industry on fire. But, that is exactly what Marvin Yoder did to the nuclear industry in America. The nuclear industry uses a business model developed decades ago and clings to those practices today. Nuclear technology is proven and safe but it is essentially unchanged since its inception. As is often the case, a person outside the industry has more chance of being a change agent to business models and technology choices than an insider.

This is Marvin's story of how he became the spark that ignited the creative juices of the nuclear industry and propelled it into a new direction. As City Manager of a tiny community in Alaska, Marvin was an unlikely person to change an industry. His strategic thinking and actions in the Galena Alaska project made the name Galena synonymous with small reactors and the possibility of new thinking in the stick in the mud nuclear industry.

When I met Marvin and I was immediately impressed with his drive, knowledge of the energy industry, Alaska connections and pragmatism. When he's on a mission, such as supplying an energy source for the community, Marvin is fearless and driven by first principles; defining problems, discussing solutions, making decisions and moving ahead. With Marvin in the lead of a team of other nuclear visionaries we became allies in "what needs to be done next" in the Galena SMR campaign.

While it was clear an Alaska nuclear project was in its embryonic stages, Marvin's vision, tactics and strategy were inspirational then and remain so today. This book tells the Galena small reactor project story and in an engaging style. It chronicles the journey of a leader and a disruptive technology in a quest to move the energy supply discussion to a new level. This story is not over and events are still in motion that are the results of Marvin's leadership. So while this is just a story of the beginning, it is a must read for anyone who wants to learn how to change the status quo.

Philip O. Moor P.E.

Chairman – ANS President's Special Committee on SMR Generic Licensing Issues

ACKNOWLEDGEMENTS

There were many people involved in providing information, offering suggestions and encouragement and in showing general interest in the Galena Nuclear proposal. I can not name them all but I do need to acknowledge the following people.

Mayor Russ Sweetsir and Vice Mayor Tom Johnson and the rest of the City Council continually placed the best interests of the community ahead of their own interests. Throughout this project, they sought additional information and made decisions based on the facts presented to them.

The Law firm of Shaw Pittman patiently listened to our questions and offered their time and knowledge to us. In particular, Doug Rosinski, Charles Peterson and Matias Travieso-Diaz took personal interest in Galena.

The Engineering firm of Burns & Roe who were also open to our inquiries. Bart Roe, Philip Moor and Charles Hess were extremely helpful during the learning phase. A special thanks to Philip who encouraged me in the writing of this book, made some technical corrections and reviewed the manuscript. I couldn't have completed the book with out him.

The Board of Directors of the Alaska Power Association who showed keen interest in the Galena project and requested numerous updates. Individual members provided feedback at various junctures along the way.

The State Administration along with the Alaska State Legislature showed their interest by funding the contract for the white papers and by making changes to Alaska State Statutes.

Gwen Holdmann and others at the University of Alaska Fairbanks for taking a second look at the potential for Nuclear after Galena was no longer engaged in the project.

Finally, Bob Woehl and the Cameron Group, who expanded the vision for this technology beyond Alaska.

CHAPTER 1

Galena, Alaska

Galena, Alaska, is located about 270 miles west of Fairbanks and 300 miles northwest of Anchorage. The community is on the banks of the Yukon River. The Yukon River flows through interior Alaska with the headwaters in Canada.

Located in interior Alaska, near the 66th parallel north, Galena experiences extreme cold in winter and warm temperatures in the summer. The average high and low temperatures are negative two and negative eighteen in January and sixty-eight and fifty-one in July. Extremes are negative sixty to mid-eighties.

Being located near the Arctic Circle affects the amount of daylight per day. At the winter solstice, the sun rises just slightly above the southern horizon at 11:30 am and sets in the southwest at 3:15 pm. From then until the summer solstice, they gain about six minutes of daylight per day so that on June 21 the sun dips slightly below the northern horizon at 1:20 am and rises in the northeast at 3:37 am.

Personally, the hardest month to adjust to is April, with clear days with more than twelve hours of daylight per day and the temperatures still falling below zero at night; I'd ask myself how a day that looks this great can be this cold.

At the time of statehood, in 1959, there were numerous small villages along the Yukon River. Living conditions were hard, and there was limited opportunity for cash income. By relying on the land and the river for food, they got by with limited need for a cash economy. In the winter, trapping supplied some cash with the furs sold in Fairbanks after breakup.

In summer many folks went to their family fish camp. A fish camp was sited at a location where they could put a fish wheel in the river to catch salmon. The fish wheel turned with flow of the river and was constructed with baskets that would scoop up salmon as they swam up stream. At the camp, they would smoke or dry the fish, pick berries, raise gardens, and in general, prepare food for the winter. When fish were abundant, the Alaska Fish and Game would open a commercial season, and fish were sold to fish processors. This brought in more cash to the residents.

Yukon River king salmon were highly sought after by processors looking to make lox or for smoking because of their high oil content. Alaska natives had their own process, which was a cold smoking process, and there was some commerce with those products as well. In addition to the king salmon, the Yukon also had chum and silver salmon. Besides salmon, there was northern pike, sheefish, whitefish, char, and grayling.

There were numerous animals that provided food as well, including moose, caribou, bear, beaver, and rabbit.

* * *

There are three divisions of Native Americans who reside along the Yukon River. The Athabascan Gwich'en resides in eastern Alaska and the Yukon Territory in Canada. The Athabascan Koyukon lives in the middle Yukon, and the Yup'ik Eskimos near the mouth where the Yukon River enters the Bering Sea.

The US Department of Agriculture established the Rural Development Council program to meet the needs of rural America. Forty states established these councils. It soon became evident that rural America was not the same as rural Alaska. Therefore, the Department of Agriculture designated the remote areas of Alaska as "frontier." In Alaska, the areas with no road access and no connection to a power grid are called Bush Alaska.

* * *

Galena was not established as a traditional village. In the early nineteen hundreds an iron sulfide (galena) mine began across the river. Miners came to work the mine and set up living quarters in what is now Galena. Additionally, some took advantage of the opportunity to provide wood to fire the boilers for the steamboats that traveled the river.

The traditional village of Louden was about ten miles upriver, and the villagers used the Galena area as a fish camp. Slowly, a small settlement formed in Galena, and eventually, Louden villagers moved to the new location and Galena became the home to the Louden native community.

* * *

Prior to the attack on Pearl Harbor, the United States was not directly involved in WWII. However, months earlier, in March of 1941, the Congress passed the lend-lease legislation, which directed the United States to provide military supplies to the allies, including Russia.

A large airport was constructed in Galena to provide a fueling stop for aircraft prior to crossing the Bering Sea. Later the US Air Force took over the airfield and established a forward operating base. During the Cold War, this base was very active. Their mission was to intercept Russian or other military aircraft flying in Alaska airspace over the Bering Sea.

With jobs available in Galena, people from the surrounding areas moved to Galena to live. Some air force personnel chose to stay in Galena after their tour of duty ended. They married locally and raised families there. Although Alaska natives are a majority in Galena, the percentage was less than other nearby communities'.

The Bureau of Indian Affairs (BIA) was responsible for medical care and education. In 1930, the BIA built a school in Galena, which operated until 1969. Medical services were also provided.

For many years, there were fighter jets based at the Galena air base. In later years, there were two F-16s permanently stationed here. Any report that an unknown plane was in the US airspace would see the Galena-based planes scramble, fly to the coast, intercept, and identify the aircraft.

When the Cold War was over, Congress began to look at reducing the military expense. The Base Realignment and Closure Act (BRAC) was established to allow the Congress to consider bases for closure. The Galena air base was placed on the list in the 1993 BRAC. The final decision was to reduce the mission, not close it. The active military personnel and the F-16s moved to the Elmendorf AFB in Anchorage. A private contractor was selected to maintain the base facilities.

This affected the population in Galena. With two hundred air force personnel leaving the city, the population dropped from 806 in 1990 to 675 in 2000 and even lower to 470 in 2010.

* * *

Living in interior Alaska is not for the fainthearted. Modern infrastructure is sporadic in Bush Alaska. There are communities that still lack basic plumbing. Millions are being spent each year, but it will still take many years to meet the need. Other infrastructure was also lacking. In 1984, I was traveling in the Norton Sound area near Nome. In the city of Golovin, I needed to make a phone call. There was only a single phone in the village. To call long distance, you needed to call collect.

Transportation was difficult. Most villages did not have a regular connection to major shopping except by air since there was no road connection. Passengers arrived in their community by air, and the freight came by barge in the summer and by air in the winter. Because of this, goods and services were very expensive. A gallon of milk, because it was flown in, cost over ten dollars. Fresh fruits and vegetables were just as expensive and low quality; sometimes they were exposed to extreme cold during transportation.

Galena was lacking in many essential services. Piped water only served a small portion of the town. Most of the residents had water tanks in their homes and either hauled their own water or paid the city to deliver water.

Sewerage was another issue. The city offices and the school were connected directly to the sewer lagoon with gravity drain pipes through the utilidors. A sewer collection system had been built in the eighties, but the system failed. Most residents had sewer holding tanks, which the city emptied with a vacuum truck. Others used "honey buckets," which was a five-gallon bucket lined with a plastic bag and a toilet seat on the top. When it was full, the homeowner took the bucket and dumped it in the lagoon. By the nineties, Galena had eliminated most outhouses.

In an area where winter temperatures dropped to fifty degrees below zero and the ground was frozen, installing piped water and sewer created immense challenges. There were programs from the federal government that could provide assistance to construct these systems. Construction cost was high, and there were many communities that needed help. These programs took years to achieve real progress.

One thing the community had to consider was that although funds to construct a new system may be available, funds for operating the system would come from the residents. Communities had to weigh whether these projects could be sustained after the project was completed.

Building a modern water treatment facility or other infrastructure increased the demand for electricity. Piped water had to be heated and constantly circulated through insulated pipes to make sure it did not freeze. This was energy intensive.

The air force had district heat, piped water, and sewer for all its facilities, but those facilities were not located near the local population. This was not a modern system and therefore not very efficient but was well maintained.

The city of Galena was a small community but acted as a hub to the communities around them. The city operated the water, sewer, and electric utilities and also operated a health clinic and a counseling center. The health clinic had two physician assistants and a health aid on staff.

The city clinic had a dental office, and a dentist was placed there by the regional tribal organization. The medical facility also employed two mental health counselors.

The other communities had paraprofessionals for both health and mental health. They consulted with, and to some extent, worked under the Galena staff. Galena health providers made regular trips to other communities.

The state of Alaska maintained three offices in Galena. The troopers had a post with two officers, and social services had two staff; both offices served the surrounding area. The Alaska Department of Transportation and Public Facilities maintained the airport and some roads. In addition, the US Fish and Wildlife Service had an office, which managed the Koyukuk and the Nowitna National Wildlife Refuges.

In 1996, I was hired to be the city manager of Galena. It was clear that the priorities from the council were to address the need for additional infrastructure. Further, the loss of the air force personnel along with base downsizing created economic issues that needed to be considered. The community was looking for ways to move the community into the twenty-first century.

CHAPTER 2

Infrastructure Needs

Galena was a small community but was very dynamic. It was important to get a community consensus on the infrastructure needs of the city. In October of 1996, the city hired Dick LaFever, an outside consultant, to facilitate an all-day community-visioning meeting to discuss the current and future needs of the community.

The results of the community-visioning session formed the basis for the completion of an updated Galena comprehensive plan. City council members, Louden tribal members, school board members, air force personnel, and the public participated in setting community goals and objectives. There was good participation from the community, and many ideas were shared.

The top priority was infrastructure improvements including water, sewer, and electricity. These were all utilities operated by the city of Galena.

Because of the downsizing and reduction of forces, several of the air force facilities were now vacant. There was a consensus that, if available, the facilities could provide economic benefit to the community. The community-visioning session was a way to receive input from the citizens regarding a potential use of these facilities. The community made it clear that the city should seek the use of the vacant air force facilities to expand education opportunities.

There are three types of school districts in Alaska. Alaska has boroughs instead of counties. All boroughs have education powers, and they oversee the operation of the school district through an elected school board. However, there are areas of the state where no borough exists.

Areas outside of boroughs are a part of a Rural Education Attendance Area (REAA) school district. The REAAs serve large areas of the state; they have elected school boards and are financially dependent on the state of Alaska.

By a vote of the residents, Alaskan cities can choose to be a first-class city. First-class cities operate city school districts outside of boroughs. Galena voted to be a first-class city shortly after statehood specifically because they wanted the local community to be in charge of education.

The year before I became city manager, the Galena school district hired Carl Knudsen as school superintendent. Carl had retired from several similar positions in Montana. He was a person who set ambitious goals and had the tenacity to accomplish the goals he set.

When there was cooperation between a city and the school district administration, a certain synergy ensued, which enabled the school district to expand their education goals. The positive working relationship between my office and the school superintendent certainly contributed to attainment of the community goals.

The community consensus was to use the vacant facilities on the base to expand education opportunities, specifically a boarding school. The city began negotiations with the air force to lease several buildings. The city agreed to pay for utilities in the buildings they would be using. The facilities we requested included a large dormitory, an automotive shop, and the abandoned NCO club. The city also gained a shared use in the base gymnasium.

Community visioning and an updated comprehensive plan provided the city council and the administration with a roadmap to improving the livability of the community.

* * *

The new millennium saw many changes in Galena. The city had begun to construct new water and sewer services. A Fairbanks firm, Lifewater Engineering, invented a three-stage septic treatment system suitable for use in the Arctic. The level of treatment was sufficient that both the EPA and the Alaska Department of Environmental Conservation certified the system as suitable for ground discharge. This, along with some septic tanks and leach fields, eliminated the need for a piped sewerage collection system.

Though not completed until 2002, plans were underway to construct of a new city hall and health clinic. In 1996 the health clinic, the mental health clinic, the city offices, and the police station were in separate buildings. Several of the buildings were old and had been moved more than once.

The city was able to combine all the various users into a single building. To accommodate the IT needs, the entire facility was equipped with fiber-optic cable. With high-speed internet service, the health clinic was able to use modern telemedicine and teleradiology equipments. This allowed the physician assistants to have instant contact with doctors in Fairbanks. The mental health clinic had a video connection, which enabled our counselors to have face-to-face interaction between a client and a psychiatrist in Anchorage.

The Galena City school district had begun operating a boarding school using the underutilized air base facilities. Enrollment on base was about 120 students in addition to 120 students in the local school.

In 1998 the district began a distance learning / home school program with approximately one thousand students. By 2000 this had grown to almost three thousand students, and the district had home-school resource centers in four Alaskan cities.

As it turned out, many of the homeschooled students were active military. In order to continue to serve these students, the district obtained a contract with the Department of Defense to provide homeschooling assistance for students when the families were deployed to bases outside of the United States.

The school district also developed a vocational program. Emphasizing vocations appropriate for rural Alaska, they had vocational studies in auto and small engine mechanics, cosmetology, culinary arts, and aviation. These were available for both high school students and postsecondary students.

This expansion required several buildings in addition to the unused air force facilities. The city leased a hangar on the airport flight line for aviation training and, in cooperation with the University of Alaska, secured a twin-engine flight simulator.

The abandoned air force automotive shop was enlarged to include a classroom as well as a hands-on automotive shop. General Motors collaborated with Suzuki to provide about 500,000 dollars' worth of tools, automobile engines, and small recreational vehicle engines. A new facility was constructed for cosmetology.

The old NCO club became a dining hall and a classroom for cooking classes.

The goals established by the community at the visioning session were being accomplished. However, one goal remained unresolved. Finding a long-term solution for the high cost of energy remained elusive.

* * *

Galena produced electricity using six electrical generators powered by reciprocating diesel engines. This required the purchase of about 350 to 400,000 gallons of diesel per year. The entire amount was purchased during the summer and delivered by barge because the Yukon River was frozen for eight months of the year. The cost of diesel had increased by nearly 50 percent in five years.

Local residents primarily used heating fuel or propane in their homes for their furnaces, cook stoves, clothes dryers, and water heaters. With long cold winters and rising fuel prices, the cost of energy was creating a burden for the populace.

As noted earlier, an important consideration was to find an alternative to diesel-powered electricity. An early suggestion was to evaluate the potential for coal bed methane. Although coal was not abundant in the region, there were known coal seams. There appeared to be fair possibility of finding gas. Eventually, we did some seismic soundings. The results were inconclusive and did not warrant further investigation.

There were calls to strongly consider green energy. Wind was not available for any commercial application. The state of Alaska had developed resource charts for various areas to categorize the wind resource. On a scale of 1 to 5, Galena scored a 1.

Another consideration was to look at the feasibility of getting power from the Yukon River. Hydroelectric or hydrokinetic energy depends on the drop in elevation to provide the power. The topography around Galena is very flat, which meant that the Yukon River at Galena would have a slow flow rate and low head pressure. To get energy from the river would require more and larger turbines. There was concern that these units would interfere with salmon runs. Another problem was that the Yukon is frozen for eight months per year, requiring placing the units under the ice. Furthermore, during spring breakup, ice flows exert tremendous force so that the units would most likely need to be removed until the spring breakup was over. Galena continued to seek other solutions.

As ludicrous as it sounds, in a region that has almost no daylight in the winter, we did spend some time evaluating solar. This consideration was based on the city use of waste heat from the diesel generators. Waste heat was piped from the generator house to the Galena school, the swimming pool, and the city facilities including the health clinic and city hall.

Diesel generators are only 30 percent efficient. Using jacket heat from the diesels for district heating in the winter increases the efficiency of the BTU usage to around 50 to 60 percent. Using solar power to reduce the electrical load on the diesel generators from April through October would not affect the heating needs of the city facilities. The challenge here was that since a solar panel would produce low amounts of electricity, amortization would take a long time.

The most promising alternative was a coal-fired power plant. There was an exposed coal seam on the bank of the Yukon River about twelve miles upriver from Galena. Since the coal was exposed, it was no problem to obtain a sample for evaluation. The results showed that the coal was low in sulfur and had a rating of 12,000 BTu/lb.

At the turn of the century, the US Congress was considering legislation to include loan guarantees and grants to construct several large-scale "clean-coal" plants. The design for large clean-coal plants was not new. However, at that point in time, a small clean-coal plant using this same technology was theoretical. The US Geological Service contacted Galena about a proposal to build a 5 MW clean-coal plant on the East Coast and, after testing, ship the plant to Galena by barge.

The city contacted the Alaska Energy Authority (AEA) to help evaluate this technology. AEA had large developed a sophisticated spreadsheet to model a comparison of the kilowatt cost of electricity from either diesel or coal.

The city considered that even if the kilowatt costs for generating electricity from coal were similar to the cost per kilowatt from diesel, coal would be preferred. When purchasing diesel, the money is sent outside the community. Mining coal locally would provide a positive economic benefit since much of the cost to produce and transport the coal would be paid to local workers. There would not be a large labor force since a 5 MW coal plant would only require a couple of truck loads of coal per day.

Two problems arose that all but eliminated this option. First, the energy bill in the Congress had been modified and clean coal was no longer a priority of the legislation. Secondly, and more importantly, after spending time reviewing the model, we concluded that given the small volume of coal that would be required each day, adherence to government regulations would increase the cost of the coal to more than $250 per ton. The cost per BTu was too high for consideration given the cost of fuel in 2002.

The city had reviewed numerous alternatives to diesel power, but none met the requirements of lowering the cost of electricity in the near future. We felt that we had struck out.

In May of 2003, a brief fax came to the city manager's office. It said, "Toshiba has developed a nuclear battery that can be buried underground and will provide power for thirty years." It was signed Steve.

CHAPTER 3

A Chance Meeting

In 1996, Galena solid waste was disposed of in an open landfill. The air force maintained the landfill. It did not have a permit; it was located in the Yukon River flood plain, and it was located at the end of the airport runway. There was open burning, and resulting smoke drifted toward the airport on a regular basis. Obviously, something needed to be done.

To bring this into compliance with Alaska Department of Environmental Conservation regulations, the city agreed to construct a new landfill. A site was selected about six miles from the community and at an elevation above any flood concerns.

In this effort, the city of Galena and the Louden tribal government cooperated to develop a plan and to secure the funds required to construct the landfill as well as the necessary equipment. Steve Howdeshell was the environmental coordinator for the Louden tribe.

He was a valuable part of the planning team. The city eventually constructed a transfer site in the community with a compactor. City personnel transported the compacted garbage to the landfill and buried it in trenches. After the landfill was in operation, it was noticed that every small breeze would blow plastic grocery bags into the trees around the landfill. In 1998, Steve approached the city council and requested they pass an ordinance forbidding plastic grocery bags in the retail stores. The city council passed one of the early laws that prohibited the use of plastic bags.

Steve left his position with the Louden tribe and went to work for Tanana Chiefs Conference (TCC). TCC was also a tribal organization but served the entire region rather than one community.

It was in this capacity that he had the opportunity to meet with Douglas Rosinski. Doug's recollection of their meeting is recounted here.

In the early 2000s, my work as a nuclear energy attorney included participation in several initiatives to kick-start what was then unfortunately referred to as the nuclear renaissance. As the promised "renaissance" of conventional, huge, base-loaded plants was not very evident at the time (and even now is barely noticeable), Shaw Pittman was participating in and, in some cases, leading the development of initiatives for nuclear projects based on innovative nuclear technology. Toshiba was a client of the firm and had explained its 4S nuclear battery concept to Charles Peterson, one of the senior Shaw Pittman energy attorneys.

In one of those "oh, by the way" moments that can only happen in a nuclear law firm, Peterson stopped by my office one day in mid-2002 and said that "if I ever found anyone who needed a small nuclear battery, please let him know." I said "sure" and that "he would get the first call." And then I went back to work.

Some months later, Steve Howdeshell e-mailed me about an environmental conference in Fairbanks that he had some connection with and which was looking for someone to make a presentation about the nuclear licensing process (not sure why). I knew Steve from an earlier exchange of e-mails regarding whether I could help an elderly veteran get his benefits from the VA, which was another part of my law practice. The firm approved my travel, so off I went to Fairbanks.

The conference itself was not very interesting, mostly people with an agenda pretending to listen to opposing or different views and then ignoring anyone or any position that did conform to the preconceived notions of the organizers.

Steve, however, was a great host and volunteered to take me on a drive through interior Alaska. The drive turned out to be about ten hours out and another ten hours back, with stops at several of the small communities with which Steve had worked or had some other contact over the years. It was an amazing trip for someone who grew up in Florida.

At one of the towns, I believe it was Delta Junction, we stopped to visit with the local town officials. While in the building housing their offices, I noticed the lights dimming in a way I remembered from my navy days when my submarine was running on its diesel engine. I made some comment to Steve along the lines of "you guys running the diesel today?" He looked at me like I was a complete idiot. It was then that I learned that outside of the "big cities," everyone in Alaska ran on diesel-generated electricity.

Once I assured myself he was not kidding, I asked Steve if he would like to replace his diesels with a small nuclear battery. Steve said, "I'm not that crazy . . . but I know someone who is." And so Steve made his first call to Marvin Yoder.

CHAPTER 4

The Toshiba 4S

Using nuclear power to produce electricity began in the 1950s. Within ten years, the US government was conducting research at various locations around the country. The Idaho National Laboratory was one such facility.

For the general public, there are two primary concerns when considering expanding nuclear power. The first is what happens to the nuclear waste. The second concern involves what happens when a nuclear power plant overheats and begins to melt the nuclear fuel.

The Idaho Laboratory conducted experiments that dealt with both of these concerns. The laboratory built an experimental 18 MW sodium-cooled fast reactor.

Because it was a fast reactor, it was able to use reprocessed nuclear waste as the fuel. Reprocessing the waste and using it a second time reduces the volume by 90 percent as well as the radioactivity of the remaining waste. Another part of the experiment involved testing for negative reactivity temperature coefficient. Simply meaning that if the coolant surrounding the core began to heat up, the reactor would slow down; and conversely, if the coolant was too cool, the reactor would speed up.

On two separate occasions, the reactor was ramped up to maximum output. Then the steam turbines were turned off to simulate a catastrophic loss of the electrical generation facility. In both instances, the reactor sensed the increased heat in the cooling system and shut down the reactor with no human intervention.

In 1990, the US Department of Energy ended the testing of this reactor.

* * *

Norihiko Handa, PhD, was a senior fellow with the Toshiba Nuclear Energy Systems & Services Division. Dr. Handa envisioned potential uses beyond electricity.

The Toshiba team concept was a small reactor that could supply various products. Since these units would be designed for remote areas, they would be operated by entities with limited resources. The reactor was named the Toshiba 4S—super-safe, small, and simple. This was the focus of the research effort by Toshiba: develop an off-grid source of electricity and other products.

Using sodium as a coolant allowed the reactor to operate at a much higher temperature and still have the coolant move through the system at atmospheric pressure. There were cooling loops in the reactor. One loop came into contact with the core and was radioactive. The second loop was heated by a heat exchanger, and this loop provided the usable heat. The secondary sodium loop had an inlet temperature of 590 degrees Fahrenheit and an outlet temperature of 905 degrees Fahrenheit.

At these temperatures, there are possibilities beyond making steam to power turbines. Dr Handa initially began to design the reactor to be used for water desalinization. Additionally, they found that by combining the electrical generation and the heat, they produce oxygen and hydrogen. Toshiba began to consider the marketing potential.

One of their early considerations of the use and number of units was as follows:

- Electric generation   400 units
- Desalinization   1,000 units
- Hydrogen production   1,500 units
- Industrial use   900 units

Having determined the market parameters, Toshiba began to continue the development of a small, super-safe reactor. The 4S was designed to fit a specific niche market.

Numerous features were incorporated into the development of the 4S that would make it appropriate for small isolated communities, whether for electricity, other uses, or a combination of uses.

Modular construction. The reactor would be built in a controlled environment and would be transported to the site. Once the final license was received from the Nuclear Regulatory Commission, the units could be constructed in an assembly line-type production and shipped anywhere in the world.

Reactor below grade. The reactor would be below ground level with only the generator plant and the air intakes above ground.

Liquid sodium coolant does not react with the core internals or piping; therefore, rust and other corrosion are minimized or eliminated.

Passive safety. This is extremely important. The goal was to have no systems that need to be operated in the case of an emergency. Current reactors need high volumes of water to keep the reactor cool. Failure of the cooling system creates catastrophic problems. The 4S is air-cooled through convection. Therefore, under normal operations or in an emergency, there are no fans or air pumps. The sodium is moved through the heat exchangers with electromagnetic force. There is no need for outside electrical power even if a disaster destroyed the generating facility.

Negative reactivity temperature coefficient. As noted above, this causes the reaction rate in the core to slow down as the temperature rises. The critical issue when the coolant flow is disrupted is twofold. The sodium is in liquid form. Extreme heat, around 1,000 degrees Fahrenheit, the sodium would begin to turn the liquid into a gas and increase the pressure. Above 1,100 degrees Fahrenheit, the fuel would begin to melt. Toshiba documents presented to the NRC contained a graph showing a margin of safety in the temperatures.

Thirty-year core life. Refueling could be a major cost factor because of the highly technical requirements associated with handling the reactor. This allows less technical skill requirements for twenty-nine years.

Ease of decommissioning. The concept is that at the end of thirty years, the old reactor vessel could be removed and a new one installed, which would limit or eliminate any radioactive material at the site.

Load following. Since the reaction rate changes with the temperature, the rate will increase and decrease depending on the amount of heat being used. There are no operator controls for increasing or decreasing the power. The only operator control is for manual shutdown. The chance for operator error is minimized.

Security. There are several features that will alleviate fears of nuclear proliferation in addition to those noted above. The fuel is enriched to less than 20 percent. This low quantity, non-weapons-grade uranium makes this fuel less desirable for destructive purposes. Also, since the reactor is assembled at the factory and then buried underground at the site, the opportunities to secure any of the fuel are effectively eliminated.

Dr. Handa and the Toshiba team developed a super-safe, small, and secure reactor that could operate in remote areas of the globe, with locally trained operators and minimal safety concerns.

CHAPTER 5

Is This for Real?

The fax from Steve Howdeshell elicited more questions than answers. I knew very little about nuclear power. I took one semester of chemistry in college. My score on a test regarding nuclear theory convinced me not to take a second semester. I needed more information about this nuclear battery.

I sent a couple of questions to Steve, who forwarded them to Doug Rosinski. He would send my question to Doug and then send me the response.

Steve soon got tired of being the intermediary, so he connected me with Mr. Rosinski, and we began to correspond by e-mail. Mr. Rosinski had an impressive background. He joined the navy and spent time as an operator on nuclear submarines. After he was discharged from the navy, he received a degree in nuclear engineering. He then went to law school and completed a law degree. I appreciated the fact that I could ask any question and he had a ready answer.

Things moved rapidly so that by midsummer, arrangements were finalized for representatives from Toshiba and Shaw Pittman to visit Galena and make a presentation about the Toshiba 4S reactor and power plant.

Representatives from Toshiba and Shaw Pittman law firm arrived in Galena in August 2003. Attending from Toshiba were the lead designers, Dr. Norihiko Handa and Yoshi Sakashita. Yoshi had spent some time learning English so he could interpret for us. Attending from Shaw Pittman were two attorneys Charles Peterson, the person who received the initial information from Toshiba, and Doug Rosinski.

The meeting was open to the public in Galena. The mayor and several council members were able to attend. Sidney Huntington, a native elder, and other community leaders also attended. We were pleased that Eric Yould, the then executive director of the Alaska Electrical Association, flew from Anchorage to Galena to attend the meeting.

The Toshiba presentation took about an hour. They emphasized the small size and the safety features with PowerPoint graphs and artistic renditions of the Toshiba 4S. They also had printed brochures that illustrated the design. It soon became obvious that our initial concept of a battery was misleading. Definitely, this was a small nuclear reactor that produced 10 MW (electrical megawatts) of electricity. After the presentation, we spent time touring the city and the air base.

Galena wanted to make a good impression. The city facilities were still new and modern. We had obtained an artifact from the area, a mammoth tusk that was more than eight feet long. The US Department of Fish and Wildlife Service had it polished and mounted.

We also toured the remainder of the building, including the new health clinic, and then went next door to the swimming pool.

We also showed them the school facilities on the air base with special attention to the automotive shop. We pointed out our cooperative agreement with Suzuki, memorialized by a four-by-eight-foot sign acknowledging their contribution to our vocational education program. Mr. Suzuki had visited Galena a year earlier.

After the Galena meeting, I accompanied them to Anchorage for a couple more presentations to folks interested in alternate energy sources.

* * *

Frank Murkowski was the Alaska governor from 2002 to 2006. From 1981 to 2002, Senator Murkowski was a United States senator from Alaska. From 1995 to 2001, he was chairman of the Senate Energy and Natural Resources Committee. He was familiar with the technological advances in the nuclear industry. In 2004 Toshiba was in Alaska again and had the opportunity to explain the 4S to Governor Murkowski. We then explained Galena's interest in finding an alternative to diesel generation. Governor Murkowski simply asked Toshiba how soon the state could get six of them. By the end of the meeting, we were confident of the governor's support if we decided to pursue this project.

Sarah Palin was elected governor in 2006. Prior to her election, I had the opportunity to spend some time with candidate Palin and to explain the Galena nuclear project. Her husband, Todd, was racing in the Iron Dog (a two-thousand-mile snow-machine race) and was between Nome and Galena, so she had time to listen. We were pleased when during the gubernatorial debates, a question was asked regarding nuclear power. Ms. Palin was unequivocal in her support of nuclear power. We felt of comfort knowing that the state administration was in support of Galena's investigation of this technology.

* * *

In Galena, discussions began about the potential for this technology to provide the solution to the energy problems. It is difficult to describe the emotions for the Galena leadership following the visit and presentation by Toshiba. Was this a potential answer to the community need for low-cost energy? Alternatively, was there a chance that Galena could experience a disaster?

Fortunately, both Shaw Pittman and the engineering firm of Burns and Roe were willing to provide information to the community. We were glad to listen to their advice since we knew they had no financial connection to the Toshiba 4S reactor.

When I first met Philip Moor, he was associated with Burns and Roe, a nuclear engineering firm. Philip had extensive experience in the nuclear industry along with his colleague Charles Hess. Coincidentally, both had an interest in the type of proposal Galena and Toshiba were considering.

Philip visited Anchorage in 2004 and then accompanied our team when we made the presentation to the governor.

Doug, Philip, and Charlie were a valuable resource for Galena. Since then, they have gone on to other firms. Their interest in the small reactor-value proposition remains strong. They have kept in contact with Toshiba and me even though the project does not have any momentum at this time. The project was the spark that grew to a flame as interest in the SMR industry grew exponentially from when the project was first publicized.

In 2006 the state of Alaska funded a study to produce a series of white papers analyzing the site requirements of the Toshiba 4S reactor. The focus of these papers was to lay the groundwork for Galena and other communities in Alaska and beyond. By preparing this information and making it available to other entities, they would have an advantage if they chose to consider this source of energy. The white paper considered the application of a specific technology (the Toshiba 4S) in a specific environment (Galena). The approach, done on a feasibility level, is the same as taken by NRC in considering a licensing application.

Pillsbury Winthrop Shaw Pittman Law Group (formerly Shaw Pitman) and Burns and Roe Engineering were contracted by the city to produce seven white papers on the various aspects of the Toshiba 4S. Shaw Pittman assigned Matias Travieso-Diaz to this project and Burns and Roe selected Philip Moor. The white papers covered important issues such as insurance and liability, emergency planning, decommissioning, security, safety features of containment, and seismic information. These papers are public technical documents but written for the average person to understand.

In this scenario, Galena would be a prototype. While the Toshiba 4S licensing and permitting was underway, the community would be working on a site license. Using the information we had gathered and the licensing work done by Toshiba, a second project could be constructed in a much shorter timeframe.

It is important to recognize that once the reactor unit is licensed, the cost of licensing the site and the construction of the next reactor is reduced. The site environmental studies are by their nature unique. Getting the first unit licensed and completing the first site license were some of the goals for both Galena and Toshiba. An advantage of modular construction is that an assembly line can be used to standardize construction and lower construction cost.

In the fifties and sixties, nuclear power plants were engineered and built on site. The consultant was largely on site, and every reactor was different as the industry was rapidly evolving and owners had design preferences. With the renewed interest in nuclear power, both large and small reactors began promoting specific models. Additional installations of a particular model will be uniform, depending on the site conditions.

From the mid-eighties to 2003, traditional news coverage regarding the research, progress, and innovation in the nuclear industry was almost nonexistent. The general public was therefore unaware of the advances in nuclear technology. While the public may have been left out, the Congress was not. In May of 2001, the Nuclear Regulatory Commission (NRC) sent a "Report to Congress on Small Modular Nuclear Reactors." The Toshiba 4S reactor was described in that report.

We began to understand that technology had moved far beyond the thermal light water nuclear plants that were now producing power in the United States, a.k.a. generation III technology. These small modular reactors with additional safety features were considered the "fourth generation" nuclear plants.

To move this concept beyond Galena, the entities involved—Burns and Roe, Shaw Pittman, and KBR—formed the Small Power Reactor Association (SPRA). It was the hope that this association could elevate the visibility of this technology. Spotlighting the expansion possibilities beyond Galena would provide momentum that would expedite the permitting and licensing.

Several companies paid to join the association. An executive director was hired to be the spokesperson and to promote the concept. The director would provide information to others in the electrical power and nuclear industry. However, after about a year, the association was disbanded. The

SPRA vision for deploying small modular reactors was too premature for the majority of the nuclear industry.

* * *

"Village Invited to Test Cheap, Clean Nuclear Power" was the headline in the *Anchorage Daily News* on October 21, 2003. In retrospect, this was an early story to go viral. Hundreds of press outlets, newsletters, opinion pieces, etc., reprinted the basic story and then added their own spin to the news. I received inquiries from around the world from organizations that were interested in a similar project. This was big news.

The news about the possible partnership between Galena and Toshiba became well-known. Numerous individuals with nuclear experience contacted Galena. Some offered to give us firsthand information based on their experience while others asked specific questions. Each of these questions, when answered by the experts associated with the Galena project, provided a higher degree of comfort with the potential power plant.

Galena recognized the need for an outside evaluation of nuclear power relative to other energy alternatives. In 2004 Galena worked with the US Department of Energy (DOE) office in Fairbanks to secure an independent contractor to evaluate the Toshiba proposal.

To get a comprehensive review, we asked that the study be modeled after a State of Alaska Energy Authority (AEA) study prepared in 2002. That study was a macro look at energy across the state. That report reviewed all potential alternatives, *except for nuclear*, including a look at a major expansion of transmission lines. The AEA conclusion was that for the foreseeable future, diesel would continue to be the preferred choice in the remote areas of the state. Galena asked for a microstudy for the middle Yukon that would include consideration of nuclear power.

DOE selected Science Applications International Corporation (SAIC) as the prime contractor. SAIC is a US company headquartered in McLean, Virginia, that provides government services and information technology support. Robert Chaney, from the Anchorage office, took the lead and involved three professors from the University of Alaska: Stephen G.

Colt, Ronald A. Johnson, and Richard W. Wies. Gregory J. White from the Idaho National Engineering and Environmental Laboratory also contributed.

This study gained credibility and firsthand information when representatives from Toshiba attended the 2004 Rural Alaska Energy Conference in Fairbanks. The Toshiba representatives made themselves available for one-on-one interviews with all the consultants.

The consultants evaluated three alternatives in depth: enhanced diesel, coal, and the Toshiba 4S nuclear plant. They also took a cursory examination of solar, biomass, wind, fuel cells, and coal bed methane.

The report recommendation was that Galena should "proceed with refining the 4S evaluation process in conjunction with the NRC." This was a further indication that this project warranted more consideration.

In 2005 Mohamed ElBaradei was the director general of the International Atomic Energy Agency for the United Nations. He received the Nobel Peace Prize for his efforts at nonproliferation. ElBaradei's mission was to monitor nuclear activity around the world. Even if a country was considering a nuclear power project, the United Nations needed assurance that the research did not provide the option of also developing materials that could be used for military or weapons purposes.

In November of that year, he lectured at MIT on the topic "Nuclear Technology in a Changing World: Have We Reached a Turning Point?" ElBaradei put forth the hypothesis that there were numerous small communities around the world where people were living in substandard conditions because they lacked basic infrastructure, primarily electric power. The dilemma, according to ElBaradei, was how to provide clean nuclear power without at the same time allowing unstable government access to destructive nuclear technology.

A solution, put forth in his lecture, was to manufacture small modular power plants that would not need refueling for thirty years. A stable country with access to nuclear technology would build and deploy these power plants. The same country would then retrieve the reactor at the end of the life cycle.

In his speech, he pointed out countries where the average household consumption was equal to the kilowatts required to operate a single lightbulb. Obviously, there was inadequate power for modern appliances, for sanitation, to keep food safe, or for economic advancement.

A similar proposal was included in the Global Nuclear Energy Partnership (GNEP), advanced by President George Bush in 2006. GNEP had four components, one of which included providing modular power plants to underdeveloped nations.

Another component involved reprocessing spent nuclear fuel, which was now stored at the US nuclear power plants.

All this was consistent with Galena's view of the niche for small modular reactors. To encourage that concept, on May 18, 2006, I testified before the Subcommittee on Energy and Air Quality of the Committee of Energy and Commerce of the House of Representatives. I explained the Galena proposal and requested the Congress provide funding for the GNEP to further this initiative.

There was synergy from both the International Atomic Energy Commission and the Bush administration to promote this technology beyond Galena. However, there was no urgency beyond these pronouncements. The Congress was cool to the concept of GNEP and did not fund it to a level that would provide any meaningful implementation.

Galena also had discussions with the Department of Energy, Nuclear Division. They explained to us that the Congress had budgeted one billion dollars per year to develop a "generation four" reactor. They did not have any excess funds to assist in the development of a small modular reactor.

CHAPTER 6

Using the Energy

By 2006, Galena was becoming comfortable with the Toshiba 4S safety and security issues. At this point in the process, the community attitude could be characterized as cautiously optimistic.

It was now time to look at the practical aspects of the project. The primary issue was that the amount of power produced was beyond our current and projected future electric demand.

Galena operated a diesel power plant with six generators. During extreme cold weather, the peak electrical demand was about 2 MW. In the summer, the demand dropped to 1 MW. The Toshiba 4S generated 10 MW.

The initial considerations were elementary. The amount of electricity used by consumers in Galena was low by national standards because the high cost, thirty cents per kilowatt, encouraged austerity. Most residents use heating fuel or propane for heat and for appliances in their homes. Reducing the cost per BTu for electricity to less than the cost of other sources of energy would increase electrical sales.

Institutional heating requirements would require a different resolution. The city of Galena provided waste heat from the diesel generator plant to several municipal facilities and was the primary heat source for these facilities. Only when the temperature was colder than thirty degrees below zero did these facilities require supplemental heating fuel. Much of the time, there was adequate heat from the generators to provide heat to the following:

> *Water plant.* The city delivered water to the residents through a six-inch insulated pipe water main. The ground is permafrost, so the water was circulated through the system and heated, above forty degrees, with a heat exchanger in the water treatment plant. Retuning water was reheated and recirculated.

> *The city school.* The school had 120 students K through 12 plus staff. There was a separate class building for shop classes. The utiliidor that housed the water pipe to the school was also heated.

> *Swimming pool.* The city built a six-lane swimming pool with upstairs recreation facilities. The building and the pool were heated by waste heat.

> *The municipal building.* This facility was heated with waste heat and included the public safety office, the city offices and council chambers, the health clinic, the dental clinic, and a mental health clinic.

Consideration of an alternative source for electricity would require that a part of the feasibility would include the cost of heating all these facilities. It was estimated that without the diesel generator waste heat, these facilities would consume around one hundred thousand gallons of heating fuel annually.

The Galena air base used 60 percent of the electricity generated by the Galena Electric Utility. In addition, they used five hundred thousand gallons of diesel heating fuel per year to provide for district heating of the air base facilities. Oil-fired furnaces produced the steam for these facilities. About ten buildings were heated with steam heat piped through underground utilidors. Heat from the nuclear power plant could heat the base facilities as well.

The SAIC study considered constructing transmission lines. There were four other villages within a seventy-mile radius of Galena. Constructing a transmission line and delivering electric power to these villages would replace the diesel generators in those communities.

Added together, these options were estimated to use about 60 percent of the available output of the Toshiba 4S. Progress toward a viable project was moving forward.

* * *

Up to this point, we were only discussing additional uses of electricity. However, the energy from a small reactor such as the Toshiba 4S can also produce other products. Finding additional uses of heat would help make the project feasible.

This is an important discussion for another reason. Most electrical power plants have several units for generating electricity. This is needed because electrical demand varies throughout the day. At 7:00 am folks are up getting ready for the day, and the demand is high. The demand drops slightly during the day but increases again in the evening and then drops to a low during the night.

Therefore, the power producer starts with a base power load that is constant and operates 24-7. The producer then adds incremental power generation as the load increases. To stay efficient, the power producer wants the demand to be around 70 to 80 percent of the generating capacity on line at the time.

Since the 4S produced more power than the city used, the issue is how to maintain efficiency. This is where adding a variable heat load was essential. As noted earlier, the second heat exchanger in the 4S facility heated water to over nine hundred degrees Fahrenheit. At this temperature, it contains a lot of energy. This energy is converted to steam to spin the turbines. When the reactor is operated at full load, the return water has cooled a couple of hundred degrees.

However, if the electrical load only uses half of the heat load, we need to find another use for the remainder of the heat so that the return water is the correct temperature. In essence, under this scenario, the "load following" is reversed compared to a standard power-producing facility. The 4S operator would monitor the electric load and then determine how much heat was available for other uses.

* * *

The University of Alaska in Fairbanks had conducted research on greenhouses for garden vegetables, and they were conducting research on indoor orchards as well. An aggressive greenhouse program would use a lot of heat, provide more nutritious and fresher food, and could lower grocery costs.

A final question was whether there was potential for further economic benefit. We received information about products, such as specialty herbs or mushrooms. With low-cost heat, these could be grown in Galena and exported. Since they would be dehydrated, the cost of shipping would be low.

Other ideas were discussed. Driving from any residence in Galena to any other area in the city was less than ten miles, would electric cars be viable? Would the batteries survive in the extreme cold? Another concept was to build a fish hatchery to enhance the fish runs in the Yukon River. The Toshiba 4S auxiliary systems could be designed to produce hydrogen and/or oxygen. Would there be an opportunity to take advantage of that capability?

* * *

There was discussion about serving other communities. Perhaps an alternative would be that rather than constructing transmission lines to other villages, we would ship hydrogen, and these communities would generate electricity with fuel cells. This would expand the number of villages that could be served.

Hydrogen is considered a future energy source. However, there are technical issues that need to be resolved before hydrogen is ready for the commercial market. There has been extensive research and experimentation using hydrogen to increase miles per gallon on diesel engines. Perhaps the school could create a technology incubator to test some of these concepts.

It was apparent that there were sufficient opportunities to use the heat and electricity from the Toshiba 4S small reactor.

* * *

I would be remiss if I did not add a final thought to the discussion about hydrogen.

In 2006, Galena received one more proposal for an energy alternative. Anchorage resident Richard Peterson is well versed in Fischer-Tropsch (F-T) technology. This technology can use coal, natural gas, or biomass to produce refined fuels. The US government encourages this process because these materials, when reduced to their basic elements (hydrogen, carbon, and oxygen) and reconstituted into gasoline or diesel, would not produce benzene, sulfur, etc. Therefore, although the fuel has the same performance characteristics, it is cleaner than gas or diesel refined from crude oil.

Mr. Peterson's primary interest was to pursue the possibility of using F-T to convert a portion of Alaska's massive volume of natural gas into refined fuel. His interest in Galena was using the forest in the area to build a biomass-to-liquids plant. Mr. Peterson had a connection to the Sasol Company, located in South Africa. Sasol is one of the largest F-T companies in the world.

At the time Mr. Peterson met with Galena officials, the pursuit of the Toshiba 4S was well underway. We discussed the pros and cons of these competing scenarios. At one point in the discussion, it was pointed out that the 4S could produce 1,624,134 scf/day of hydrogen and half that much oxygen. Mr. Peterson asked for time to confer with Sasol. The response he received added a new dimension to potential uses of the heat from the 4S. Mr. Peterson provided the following response:

> "The hydrogen can combine with the long chain F-T wax to produce more diesel molecules. The oxygen would just replace oxygen coming from the air separation plant. The hydrogen

would produce more fuels whereas the oxygen would reduce the operating costs of the BTL air separation plant. To the extent you have minimal electric loads 6 to 7 months out of the year you operate the nuclear plant at a constant load and either make electricity or Hydrogen and Oxygen. The more hydrogen the more transport fuel you make, additional oxygen allows you to run the air separation plant less. Hydrogen makes up to ten percent more fuel. Oxygen reduces your operating cost."

This concept speaks to the versatility of this technology. The ability to produce hydrogen, fresh water, or heat makes these units desirable for developing economies.

CHAPTER 7

Nuclear Concerns

All this is not to say there was no opposition. Some people are opposed to nuclear power under any circumstances. The most memorable demonstration of splitting the atom was a bomb with a devastating loss of life. That image is hard to ignore.

The cities of Hiroshima and Nagasaki, which were actually bombed by atomic weapons, have cleaned up the residual contamination and thrived. Studies abound on the cleanup and chronic and long-term health effects on victims of those bombings. Moreover, because of the close proximity of time, there are folks who confuse the facts of the Three Mile Island incident with the movie *The China Syndrome*. It must be noted that there were no direct fatalities or any long-term environmental degradation connected to the Three Mile Island incident.

The United States has been producing electricity from nuclear power plants for sixty years. Considering the advanced technological nature of this science, the safety record of the United States nuclear power is remarkable. The statistics from the Occupational Safety and Health Administration (OSHA) show that nuclear power plants have the best safety record of all the electric power producers.

This safety record is attributable in part to the rigorous process required by the Nuclear Regulatory Commission before a license to operate is granted.

We began to learn about the process required for Toshiba 4S to obtain a license being onerous. First, the company sponsoring the design must complete massive amounts of research and development to produce a sound scientific basis for the project. Then components of the proposed unit are tested to verify the scientific theory. Toshiba then presents all their information, both the science and the full-scale model test results to the NRC. Toshiba then must pay the NRC to review the information and to verify the test results. The staff at the NRC then reviews the information from Toshiba and the NRC's consultants.

* * *

Some residents were hesitant to embrace this project; however, there was not much vocal opposition in Galena. On the other hand, the Yukon River Inter-Tribal Watershed Council (YRITWC) immediately stated their opposition to any nuclear power plant on the Yukon River.

YRITWC was an advocacy group, active on the Yukon River in both Canada and Alaska. Their stated goal was to "make the Yukon River clean enough to drink." This seemed somewhat opportunistic since the Yukon has no industry or agriculture along its banks. There is only minimal opportunity for chemicals, pesticides, or fertilizers to leach into the river.

Their concept of nuclear power was based on conventional thermal reactors. So a part of their opposition was a misunderstanding of the technology engineered into the Toshiba 4S, which is an evolutionary design. One of their stated complaints was that cooling for a nuclear reactor would add heat to the Yukon River, which would change the river's characteristics. The amount of heat required to heat a river as enormous as the Yukon is millions of times that rejected by a small reactor. However, there was fundamental disconnect; the 4S is air-cooled and therefore would not impact the river.

Another concern that was vocalized by their spokesperson was that Toshiba was looking for an isolated spot where any problems would not be noticed by the rest of the world. Of course, that ignores the role of the NRC in locating and licensing a nuclear plant. In fact, the motivation for Toshiba to place a facility in the United States was that they considered the NRC to be the gold standard for nuclear facility licensing. A license by the NRC opened the door to market the 4S to the world.

One point seemed to be forgotten. Annually, shipping over 1,400,000 gallons of fuel on the river carried its own risks. Compare that to the rather low probability risk of a problem with the highly evolved technology of a modular reactor.

YRITWC also commented that we should consider alternatives and said they had the expertise and wherewithal to assist. Galena coordinated with YRITWC in our previous energy evaluations, but nothing seemed viable. In 2009 they conducted a short demonstration project for hydrokinetic power in the village of Ruby.

Even with some outside opposition, the city council on two occasions approved a resolution to pursue the nuclear project. The first resolution was passed in 2004, and since the resolutions had a two-year sunset clause, the resolution was passed again in 2006.

The resolution stated that the city councils determined it was in the public interest to pursue the sitting of a nuclear plant in Galena and directed the city manager to establish a process and a timeline leading to the evaluation, industrial partners, and financial and contractual arrangements necessary to bring the economic and environmental benefits to Galena.

* * *

Japan suffered a disastrous earthquake and tsunami on March 11, 2011. There were several operating nuclear power plants, each with multiple reactors, in the devastated area. All survived the initial earthquake, and all except Fukushima Daiichi maintained backup electricity and shut down normally.

Fukushima Daiichi lost power as a result of the earthquake and its twelve backup generators as a result of the Tsunami. Without power, it was not possible to circulate the cooling water. With inadequate cooling, the reactor began to melt down. For many days there were images in the press of the workers attempting to use any type of water pump available to try to get cooling water on the reactor. It was an extremely tense time.

The earthquake and tsunami caused about nineteen thousand deaths in the region. Three of workers at the power company lost their lives in the tsunami. None of the workers lost their lives from radiation. Fukushima Daiichi is similar to the thermal reactor plants now in the United States.

The advantage of the evolutionary design in sodium-cooled nuclear reactors is that during a catastrophe, additional cooling is not required since the new plants rely on negative reactivity coefficient. The Toshiba 4S and other reactors of this type will not need electricity or pumps in the event of catastrophic loss of the generating system.

* * *

The city had received massive amounts of information. It was felt that the Toshiba 4S had been designed with the safety and security features and the simplicity to make it suitable for a small isolated community.

CHAPTER 8

Meeting Reality

An optimist by nature, I enjoyed two quotes during my time as manager:

> *We are continually faced with a series of Great Opportunities brilliantly disguised as Insoluble Problems.* (John W. Gardner)

Another quote:

> *The difficult we do right away, the impossible takes a little longer.*

So the question is, Was it hubris or naiveté that convinced us that we could complete this project? The complexity of a "first of a kind" nuclear reactor was beyond what we anticipated. The challenges were numerous.

Tom Johnson, Galena vice mayor, and I met the NRC staff in 2004 and requested that they consider approving the Toshiba 4S for Galena. The Nuclear Regulatory Commission had not licensed a new nuclear power plant since 1980. Furthermore, this was the first time a municipality came to them asking for a nuclear power plant. Additionally, the request was for a nontraditional design. All the licensed nuclear power plants operating in the United States were light water reactors; the Toshiba plant was designed to use liquid sodium.

It was obvious that the NRC was caught off guard with the Galena request. They began to look around the room to determine if anyone was aware of whether there were engineers still with the agency who had experience with the EBR-1 and EBR-2 sodium reactors at the Idaho National Laboratory and other sodium reactors like FFTF at Hanford. There was uncertainty on how the agency would proceed toward permitting this technology. It should be noted that when Toshiba presented their plans to the NRC in 2008, they stated that 90 percent of the design conformed to reactors that were already approved. Nontraditional design content (meaning not like light water reactors but consistent with sodium reactor experience) was approximately 10 percent.

Other issues surfaced as well. It was not realistic to think that a small community had the resources to own and operate a nuclear power plant. Galena needed to find a sponsor or an independent power producer (IPP), which would own and operate the 4S under a purchase power agreement. Galena would purchase wholesale power from the plant operator and sell retail to the customers. The IPP would also set a price per British thermal units (BTu) for heat. The city could purchase BTu's either for resale or for their own use, or the IPP could sell BTu's directly to a final user. We had preliminary discussions about entities that could qualify as a potential IPP, including the state of Alaska. However, until the NRC process was better defined, it was premature to begin negotiations.

Toshiba and the Japanese research community spent hundreds of millions of dollars on research and development. Toshiba's plan was to secure NRC approval and licensing to build an initial unit and then market these units around the world. It could cost $100 million to get the first unit 10 MW licensed. Before making that commitment, Toshiba needed a clear path forward from Galena, the NRC, and an IPP.

Galena needed to identify an IPP that would contribute to the licensing and building process. The IPP would also have substantial upfront cost during the licensing process. They would require some level of comfort that Galena could meet their obligation to purchase a large percentage of the electricity and heat. A typical IPP works with 100 percent of the output of a power plant in calculating the project economics. The Galena project would not be financially feasible if the sales were only 50 percent of the capacity.

As noted, to use the power, community residents would begin the transition from diesel and propane to electricity. The timeline for this would depend on individual actions. Galena could not expect that everyone would change over immediately, but rather there would be a gradual shift. The city would need to construct upgrades to distribution lines and to the transformers. At the same time, Galena would be expected to show progress in securing the funds to build the downriver transmission line extension as well as facilitating the greenhouse project. How these iterations of multiple parties would come together became more and more thorny.

In the summer of 2006, Galena was cautiously working toward solving all these issues to facilitate the installation of the 4S. At that point, an additional uncertainty was added to the mix. The Galena air base was now going through the second round of BRAC. This round would totally remove the air force from Galena. The air force used more electricity and heat than the rest of the community. Losing this load could erode the feasibility of a 10 MW power source. The demand for both heat and electricity would be greatly reduced if the air force withdrew all their presence from Galena.

The air force owned the buildings, but the land belonged to the state of Alaska. The contract with the state of Alaska stated that in the event the lease was terminated, the air force would return the land to its "original" condition. That meant all the structures would be demolished.

Galena now had three additional fronts to fight in the drive toward a viable project: finding an efficient source of energy, structuring the energy contracts, and preventing the demolition of the air force facilities. The community would need to utilize the abandoned air force facilities to maintain the local economy. The community embarked on a plan, with the cooperation of the state and the air force, that allowed Galena to take over ownership of a number of the air force facilities. The boarding school expanded their programs to make use of the additional space. However, some buildings were demolished and the demand for electric power declined.

On the nuclear front, there were other issues. The news reports in 2003 about Galena looking to build a nuclear power plant went around the world. Hundreds of periodicals, newsletters, and such quoted portions of the original article. It seemed nuclear power was once again a part of the discussion regarding the energy mix. Some large and small power plants were being actively investigated.

Policy makers and regulators were just beginning to realize how much nuclear power technology had advanced during unofficial moratorium. We already felt that Toshiba's R&D was ready to move the next step. There was concern that if other units, more either traditional or larger units, requested licensing, the "mini" 4S would be pushed to back of the line. Toshiba needed to be first to present a license application to the NRC.

In December 2006, Philip Moor and I went to Japan to meet with Toshiba. Our mission was to convince Toshiba that if they wanted a timely review of 4S reactor by the NRC, they should start the process within three months. If they were first, they had a better chance of success.

Unfortunately, the vice president in charge of the nuclear division for Toshiba had unexpectedly left Japan. During the early part of the discussion, it seemed like they were just buying time. Even when the Toshiba vice president returned, the negotiations were still stalled. Philip and I were preparing for an early departure from Tokyo. At the last minute, we received a call, and we had some productive discussions; however, although we left with an MOU, we did not get all the wished-for assurances.

It was not until later we learned that Toshiba was finalizing the acquisition of Westinghouse Nuclear, which prompted the vice president's absence. In 2007, when the 4S was on the NRC docket, Westinghouse was in charge of presenting the case. Westinghouse/Toshiba made four preapplication presentations to the NRC in 2007 and 2008. These were informational proceedings; Toshiba did not request a permit or a license to operate as Toshiba expected an IPP, utility or equivalent, to be the applicant for an NRC license. The presentations were to fill in the knowledge gaps because of the nontraditional design and to clarify the permitting procedures preferred by the NRC.

Clearly, the missing element at these presentations was the identification of an IPP. I continued to testify to Galena's interest based on the city council resolution. There are two required conditions before the NRC will open a docket for a permit: there must be an owner identified and the costs associated with licensing must be committed. These were not in place at that time. By the end of 2008, the NRC had other applications that took precedence.

I had retired from the city of Galena at the end of 2006. I maintained contact with Westinghouse and Toshiba, although by the end of 2008, it was clear the probability of a 4S for Galena was very low. Other small modular reactors, however, were being considered, particularly light water designs form B&W, NuScale, and others.

Additionally, the strategic value of certain types of small reactor began to receive attention. Most of the SMRs being considered were light water-cooled thermal reactors, not fast reactors. The fast reactors, like the Toshiba 4S, can use reprocessed nuclear waste. Used nuclear fuel can be reprocessed to extract fissile materials for recycling and to reduce the volume of high-level wastes. Reprocessing technologies in conjunction with fast neutron reactors will burn all long-lived actinides.

The United States unilaterally stopped developing reprocessing of nuclear waste in the 1970s. There was a concern that in the reprocessing phase, separation of fissile material for weapons could also take place.

However, other countries in Europe and Japan and Russia are currently reprocessing their waste material. The technology is available. Reasons for not doing this are primarily political.

It is necessary to understand the scope of the problem. The thermal reactors now in service in the United States are not configured to use reprocessed fuel. The used fuel is stored on-site or ultimately in a federal repository like Yucca Mountain. If they could, reprocessing all current waste now in storage would meet the fuel demand in the United States for many decades.

Storage of the waste has been an ongoing problem. Reprocessing will reduce the volume of waste by 80 percent and, by burning the actinides, reduce the radioactivity as well. Therefore, placing several hundred fast reactors around the world, using reprocessed fuel, would be a start at solving the nuclear waste storage problem.

Throughout the years, from 2003 to 2008, we were focused on how to use an SMR to lower the cost of energy for Galena. Additionally, we considered various options to use the additional heat.

In 2009, I received a telephone call from Robert Woehl, who was associated with the Cameron Group. The Cameron Group is an association of individuals who had successful careers, retired and then banded together to find solutions to new challenges.

Mr. Woehl and I spoke several times on the phone, and then I flew to San Francisco to meet with the team. Their concept was that we should consider the possibility of a sustainable community. If electric BTu's were reasonably priced, would it be possible to lower the cost of living, increase the quality of life, and eliminate the use of fossil fuels? Was it realistic to expect that we could locally produce a large percentage of the food required in Galena and create economic opportunity?

To holistically look at this technology in the global context embodied the principles presented by Mohamed ElBaradei and President Bush's global nuclear energy initiative.

- Mohamed ElBaradei, at the David J. Rose Lecture on Nuclear Technology at MIT on November 3, 2005, challenged the atomic energy producers to export a 4S or similar technology to underdeveloped countries around the world. He envisioned modular construction of nuclear power plants that would be on-site for thirty years and then be returned to the manufacturing country.

- In 2006, President Bush put forth the Global Nuclear Energy Partnership (GNEP). One component was to place SMRs in the developing countries.

Perhaps Galena could have become a demonstration of using an SMR to develop a sustainable community. However, the challenges were too great; by 2009, that train had passed by Galena. It would be up to another community to determine if that vision would meet their needs.

In the final analysis, consideration of the Toshiba 4S expanded the nuclear conversation into three areas:

1. Providing a competitive cost for electric power in isolated communities as well as isolated industrial sites such as mines

2. Creating an avenue to reduce the volume and volatility of nuclear waste

3. Providing added benefit by using the excess heat for additional benefit

When the right political and contractual elements are in play, this technology will increase health and quality of life in isolated communities around the world.

CHAPTER 9

Final Thoughts

The Toshiba/Westinghouse nuclear program is alive and well. The flagship of their products is the Westinghouse AP1000 producing over 1,100 MW. This is a generation III+ reactor with the newer safety features and a standard design. According to their website, Westinghouse has four of these units under construction in China, and five US utilities have selected this design for future construction. AP1000 units are under construction in Georgia and South Carolina.

Westinghouse has abandoned their light water SMR, and currently the plans for the Toshiba 4S fast reactor have been put on the shelf and there are no further development plans at this time.

Interest in SMRs continues to build.

In 2011, I worked with State Representative Craig Johnson to make some minor changes to the Alaska statutes that dealt with nuclear power. These changes were required before the AEA could include nuclear power in their work program.

A conference entitled "Building the Value Chain for Commercializing Small Modular Reactors" was held on October 18 to 20, 2010, in Washington DC. I was asked to be a presenter on one of the panels.

The University of Alaska Fairbanks (UAF) sent representatives to that conference. The university realized that answers to the electrical needs in Bush Alaska had no easy answers, and they were intrigued by the possibility of using nuclear energy for remote heat and power applications. Gwen Holdmann, who directs the Alaska Center for Energy and Power (ACEP) at UAF, led the group.

Prior to Gwen's current assignment at the university, she had spent time working at Chena Hot Springs, northeast of Fairbanks. As the engineer tasked with making this remote, off-grid site energy self-sufficient, she focused on finding a way to use the hot water from the hot springs to make electricity. Most turbines operate with steam, which condenses after spinning a turbine. However, the water at Chena Hot Springs was less than 165 degrees Fahrenheit. Gwen was asked to find a way to generate electricity at this lower temperature. Today, this is the lowest operating commercial geothermal power plant in the world.

After that project was completed, Gwen was hired by the University of Alaska Fairbanks (UAF). She was interested in continuing the nuclear discussion for Alaska as a way to bring reliable, affordable energy to remote communities that do not necessarily have easy access to renewable resources such as geothermal. She and a couple of colleagues attended the conference to learn more.

Then university and the Alaska Energy Authority (AEA) cooperated to bring a nuclear energy workshop in Alaska. They also edited a chapter on the potential for nuclear power in Alaska in the energy handbook produced by the AEA.

The ACEP produced a final report entitled "Small Scale Modular Nuclear Power: An Option for Alaska." This is available on their website. Gwen and her colleagues continue to believe that small modular nuclear energy could have a future application in Alaska, and they are actively tracking developments in the industry.

* * *

The state of South Carolina decided to take the lead in promoting SMRs. The state expressed a willingness to work with companies that were planning to build these reactors.

The first annual SMR conference was held in Charlotte, South Carolina, on April of 2011. There were several hundred participants from the nuclear community. Also attending the conference were legislators from the state and the US Congress. The elected officials were among those who made presentations. I was invited to make a presentation about lessons learned from the Galena project.

In 2004, the timeline developed for Galena projected that construction would begin in 2011. At the conference, there were speakers who were optimistically predicting that an SMR could be licensed by 2020. It occurred to me that if the GNEP vision was to become reality, a revised timeline was needed. Unless some urgency was added to the NRC licensing process, the United States would be importing rather than exporting this technology.

At the conference, there was a sense of anticipation created when it was announced that the Department of Energy (DOE) would be accepting proposals from the firms developing SMRs. The winning proposal would have a cost-sharing arrangement with DOE to develop their SMR technology.

The first proposal accepted by the DOE was Babcock & Wilcox, which had experience in building reactors for navy ships. They were developing an SMR labeled mPower that would produce 180 MW. Since their plan was to build them in pairs, the completed facility would produce 360 MW of electrical generation. B&W's mPower has been in preapplication design interaction with the NRC since 2009. They intend to submit an application for design certification in the third quarter of 2014.

In December of 2013, DOE announced that NuScale was selected as the second winner of the competition. NuScale was developed at Oregon State University in Corvallis. NuScale will receive funding to support the accelerated development of their module. NuScale is developing a 45 MW module. Each installation consists of twelve 45 MW units, although they can be built with six in the initial phase. Therefore, the installation, when complete, is designed to produce 540 MW.

The mPower and the NuScale have many of the same safety features as the Toshiba 4S. However, they are light water reactors and fairly conventional except for the smaller size.

These are called small because they are compared to the Westinghouse AP1000, which produces 1,100 MW, and the GE ESBWR, which produces 1,500 MW.

The SMRs have several advantages to the large plants and can produce power at a competitive price per kilowatt. One advantage is the ability to add capacity incrementally. Adding 250 MW every three years can have a financing advantage over adding a single 1,000 MW unit in twelve years. It allows the utility to be more efficient since they won't have a lot of unused capacity.

A second issue is distributed generation. One of the looming problems in America is the capacity of the transmission lines. Placing 2,000 MW in one location puts pressure on the power distribution lines. Placing several smaller units at choke points in the grid will strengthen reliability and reduce strain on the system. Reducing the need to overhaul major transmission lines would provide substantial cost savings.

There is also a national security aspect to the idea of distributed generation. In recent months, there have been several incidents of vandalism at electrical substations around the nation. Officials are concerned that a major coordinated attack on these substations could have a disastrous effect on our economy. Distributed power can reduce the vulnerability to this threat by providing alternative routes for electric transmission.

To date, however, these SMRs have not received the expected interest. Concern for national security may be the impetus to move this technology forward. At least in the United States, this program seems stalled.

However, for purposes of this book, 200 MW is very different from 10 MW. These small modular reactors for the most part are designed to enhance and/or supplement the existing grid. The market we envisioned is off-grid in isolated areas, away from other development. These can provide energy to a community, several communities, a remote mine, or another industry.

* * *

The issues raised in this book point to a larger problem. The United States does not have a comprehensive energy policy! The Congress has debated and, in some instances, introduced legislation to develop a comprehensive, long-range policy. However, special interests get involved. Instead, they pass legislation with short-term stopgaps, such as corn ethanol, wind generators, or café standards for cars to solve the energy needs of the country. At that point, a comprehensive policy is reduced to a couple policies that may or may not produce the desired results.

The ideas put forth here will be implemented only if the United States government takes the initiative to establish a comprehensive strategy for energy in the United States and nuclear issues worldwide.

Small, or perhaps minireactors, are needed because of the special niche they fill. Remote locations, whether for a community or for industry, would, in many cases, find 100 MW or less entirely sufficient to their needs. Our goal is to enable these communities to become self-sufficient regardless if the output is electricity, clean water, heat, or hydrogen.

Galena will not demonstrate to the world the path forward. Our hope is that a roadmap was created for others to follow.

APPENDIX

Additional reading:

More information on the life of the Koyukon people, *Shadows on the Koyukuk* by Sidney Huntington

More information comparing nuclear power to all other sources, *Terrestrial Energy* by William Tucker

Available on the web:

"*Report to Congress on Small Modular Nuclear Reactors, May 01.*"

"*Galena Electric Power—a Situational Analysis.*" Institute of Social and Economic Research. www.iser.uaa.alaska.edu/ . . . /Galena_power_ University of Alaska Anch . . .

"City of Galena Presentation of *White Papers* for the *Toshiba 4S* Nuclear Reactor."

"City of Galena Comprehensive Plan." www.commerce.state.ak.us/dca/plans/Galena-CP-1998.pdf.

http://energy.gov/sites/prod/files/edg/news/archives/documents/GNEP/06-GA50035f_2-col.pdf.

ElBaradei. http://web.mit.edu/newsoffice/2005/nuclear.html.

Numerous articles from http://www.world-nuclear.org/ and http://k1project.org/hiroshima-and-nagasaki-the-long-term-health-effects/.